# Science Crossword Puzzles
## Grades 2–4

```
                                            F
                                            O
              T                             R
              E                             C
              R
    A  N  I  M  A  L  F  A  M  I  L  I  E  S
    N        S
    I
    M  A  T  T  E  R
    A
    P  L  A  N  E  T  S
    C
    H  U  M  A  N  B  O  D  Y
    A
E   N  E  R  G  Y
    A
    C  L  I  M  A  T  E
    T
    W  E  A  T  H  E  R
    R
    I
    S
    M  O  T  I  O  N
    I
    C
P  L  A  N  T  S
```

## Written by Rebecca Stark

ISBN 978-1-56644-569-6

**Educational Books 'n' Bingo**

Printed in the United States of America.

ISBN 978-1-54466-569-6

EDUCATIONAL BOOKS 'N' BINGO

Printed in the United States of America

# TABLE OF CONTENTS

*An alphabetical list of possible answers from which to choose is provided for each crossword puzzle. Use these lists at your discretion.

# Animal Characteristics

**ACROSS**

6   Animal that eats meat

7   Insect with 4 stages: egg, larva, pupa and adult

8   An animal whose body temperature doesn't change with its environment (hyphenated)

12   Animals that breathe through gills their entire lives

14   Animals that do not eat meat

15   Animals with hair

17   Seals and walruses use them to swim

19   The elongated nose of an elephant

20   A large mammal that lives in the ocean

**DOWN**

1   Known for its very long neck

2   Mice and other animals with continuously growing incisor teeth

3   Complete this analogy: dog : canine :: cat : ___

4   Like foxes and wolves, they are members of the canine family

5   A snake is one; so is a turtle

6   Animal whose body temperature changes with its environment (hyphenated)

7   Animals with feathers

9   Live part of their lives in the water

10   Female reptiles and birds lay ___ with a hard, protective shell

11   Describes a species that no longer exists

13   Humans and apes, for example

16   Fish and young amphibians breathe through them

17   An amphibian

18   Female kangaroos and other marsupials have one

# Animal Characteristics

# Animal Families

## ACROSS

1   A baby dog

5   A horse, male or female, less than one year of age

6   A young goat

8   A large number of fish of one kind swimming together

10  A male duck

12  An adult female horse

13  A baby bear

14  A joey is a baby one

16  A baby fox

18  A baby cat

20  Antelopes travel in one

## DOWN

1   A group of lions

2   A group of kangaroos

3   Birds traveling together

4   A group of baboons

7   A baby cow or buffalo

9   A baby sheep

10  A female deer

11  An adult male horse

13  A young male horse

15  A young goose

17  A young female horse

19  A female sheep

# Animal Families

# Energy

## ACROSS

4   Amount of energy in a substance

5   Magnitude of sound

6   Able to attract certain metals

8   Energy acquired by the objects upon which work is done

10  Energy from a source that is not depleted when used

12  Allows electricity or heat to flow through it

13  Flow of electricity through a conductor

15  Does not allow the electricity or heat to flow through it

16  Movement of air; can be converted into mechanical power

17  Electrical energy in nature

20  Container that stores energy

21  Act of using force to move something over a certain distance

23  Electricity produced by friction

24  Radiant energy from the sun

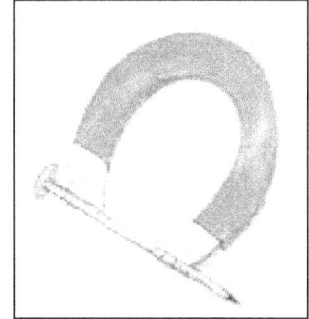

## DOWN

1   Energy that travels through air and can be heard by the ear

2   Stored energy is called ___ energy

3   Batteries, dry wood, and coal are examples of stored ___ energy

7   Needed to do work

9   Flow of electric charge through a conductor

11  Source of light energy

14  ___ energy is energy in motion

18  Energy that can be measured by using a thermometer

19  Force that tries to pull two objects toward each other

22  Energy that can be sensed by the eye

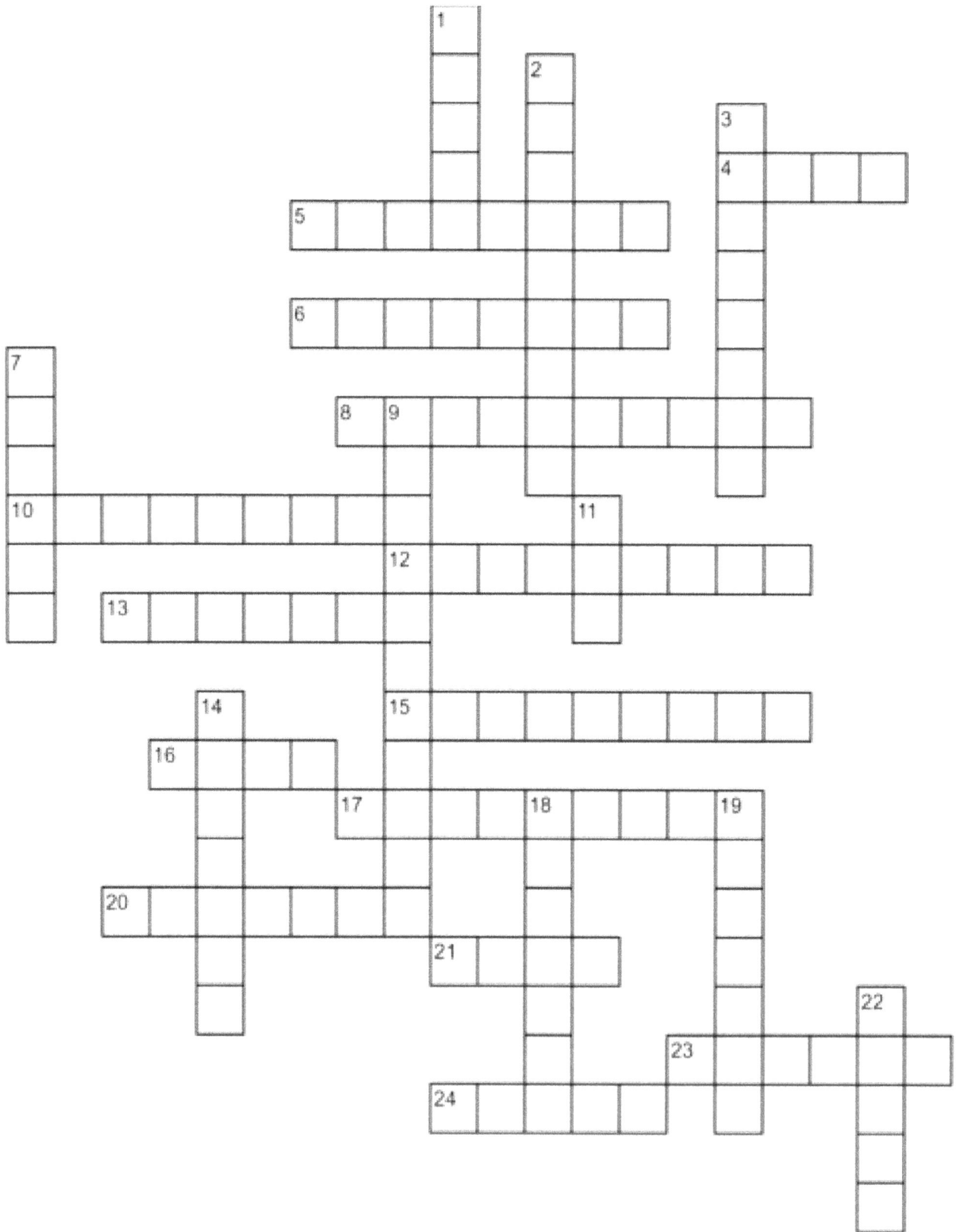

# Energy

# Force and Motion

**ACROSS**

3    Tendency not to move or change

5    A push or pull on an object

7    A wheel on a fixed axle; has a groove along the edges to guide a rope or cable

8    An object that attracts metals such as iron, nickel and cobalt

9    Force that tries to pull two objects toward each other

10   A change in position compared to a place or an object that is not moving

11   Lever, pulley, wedge, wheel and axle, screw and wedge (2 words)

12   The force of magnets

13   Actual amount of matter in an object

15   Ramp that assists moving heavy objects up and down heights (2 words)

16   It combines the force of electricity and magnetism

17   Wheel with a rod in the middle to help it to lift or move loads (3 words)

19   Path along which something moves, lies, or points

21   Energy produced by moving charged particles

22   To pull toward itself

24   To use force to move an object away from something

25   To use force to move an object towards something

**DOWN**

1    Rate at which an object covers distance

2    Simple machine with triangular shape; can be used to separate objects

4    Point on which a lever turns or balances

6    A machine made up of 2 or more simple machines

11   A stationary electrical charge (2 words)

14   A cylindrical rod with spiral threads

18   A seesaw is an example of this type of simple machine

20   How heavy something is

23   To force something to move away

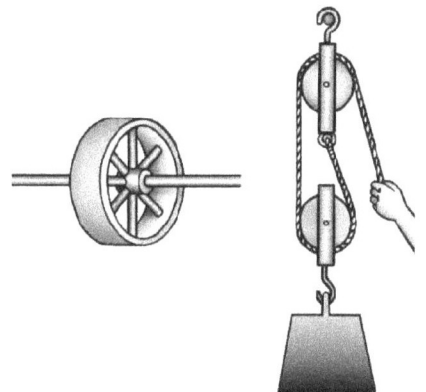

# Force and Motion

# The Human Body

## ACROSS

3   Breathing organs
4   Basic unit of any living thing
5   Contains our brain, eyes, ears, nose, and mouth
6   Forehead, eyes, cheeks, ears, nose and mouth are parts of this
9   Organ of thought; inside the skull
10  Joint that connect forearm and upper arm
12  Hand, wrist and elbow are part of it
13  Your arm connects to your body at this joint
14  Front part of leg below the knee
17  Carry signals between the brain, spinal cord, and the parts of the body
18  System that provides structure, protection, and support for the body
20  Front part of the body between the neck and stomach
22  Carries nutrients to each cell and takes away wastes
23  Support our body and protect our internal organs
24  Joint that connects hand and forearm

## DOWN

1   Protects brain; part of skeletal system
2   Joint between the thigh and the lower leg
5   Pumps blood through our body
7   Similar cells that work together to carry out a specific function
8   Sight, smell, sound, taste, and touch
11  Area above the hips
13  Bundle of nerves that runs up and down the center of the back (2 words)
15  Between your head and your shoulders
16  Section from hip to knee
18  A group of organs that work together to carry out a particular task
19  Joint that connects leg and foot
21  Where 2 bones come together
25  Outer covering of the human body; sense organ for touch

# The Human Body

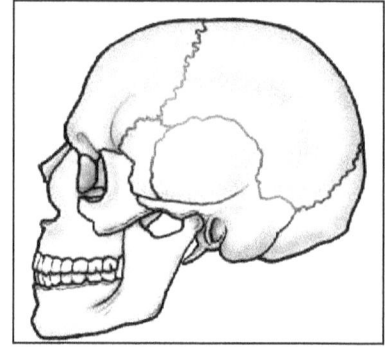

# Matter

**ACROSS**

3   To change from a solid to a liquid by adding hot energy

6   A liquid's shape depends upon the shape of its ___

8   Gaseous state of water

10   Size, shape, color, weight, and texture are ___ of matter that we can observe

13   To mix a solid with a liquid so that it becomes part of the liquid

15   Solid, liquid, and gas are the 3 ___ of matter

17   Has mass and takes up space

18   How much space matter takes up

19   A liquid in which something has been dissolved

20   A ___ change, or reaction, that results in a different kind of matter

**DOWN**

1   To change from a liquid to a solid by adding cold energy

2   A gas in the air we breathe

3   When two or more substances are combined, but each keeps its physical properties

4   State of matter that has definite mass and volume but no definite shape

5   State of matter with definite shape, mass and volume

7   Degree of hotness or coldness

9   The amount of matter in a particular space

11   Adjectives that might describe this property are *rough, smooth, sticky,* or *fuzzy*

12   Water in its solid state

14   We use our ___ to observe the physical properties of matter

16   State of matter with no definite shape, mass or volume

20   Hue, saturation and brightness are properties of ___

21   Amount of matter in an object

22   Matter is made of these tiny particles

# Matter

# Planet Earth

## ACROSS

3   Center of Earth

6   Molten rock beneath Earth's surface

7   Landmass that projects above its surroundings

8   A flowing, moving stream of water.

11  Process by which water is circulated throughout Earth and its atmosphere (2 words)

12  Complete this analogy: Earth : planet :: sun : ___

13  Uppermost layer of Earth's crust that can be dug and plowed

14  One of Earth's 7 main areas of land

15  Layer beneath the crust

16  Earth orbits around it

17  Outer layer of Earth

19  Large amount of earth and rock that moves down a steep slope

20  Process where rock is dissolved, worn away or broken down into smaller pieces

23  The three main types: igneous, sedimentary and metamorphic

24  One ___ orbits Earth


## DOWN

1   Large body of salt water

2   Slow-moving mass of ice

4   Crack in Earth's crust

5   Any of the 8 large bodies that orbit our sun

7   Gold and silver are 2 examples

9   A vent in Earth's surface; molten rock and gases sometimes escape from it

10  Shaking or sudden shock at surface of Earth

18  Alternate rising and falling of the surface of the ocean (singular)

21  Wearing away of Earth's surface by a natural process

22  Floating mass of ice that has broken away from glacier

# Planet Earth

# Plants

## ACROSS

1  Supports the plant; sucks food, minerals and water to other parts of the plant
3  Uppermost layer of earth in which plants grow
7  Plant parts from which new plants grow
9  Edible part of a plant that holds and protects the seeds
10  Small growth that later develops into a flower, leaf, or branch
11  Process of transferring pollen to other flowers
12  Gas needed for photosynthesis (2 words)
13  Insects, birds, and bats are good ones
14  A part of a tree that grows out from the trunk
16  Green plants use sunlight, carbon dioxide and ___ to produce sugar and oxygen
17  A young plant
19  They collect sunlight for photosynthesis
20  A living thing that uses sunlight to make its own food
21  Plant part that receives pollen from a visiting pollinator

## DOWN

2  Small branches
4  ___ hold plant in place; they take and store nutrients from the soil
5  Gas given off during photosynthesis and needed by humans and other animals
6  Plant stage between seed and seedling
7  Male part that produces pollen
8  Organism, such as a plant, that makes its own food
9  Part of a plant that produces seeds
11  The process by which green plants make food
15  A fine powder that causes some plants to form seeds
18  Parts of the flower that attract pollinators with their shape, color and smell
22  The main woody stem of a tree

# Plants

# Weather and Climate

**ACROSS**

3   A disturbance in the atmosphere; hurricanes, tornadoes and blizzards are severe ones

6   Circulation of water on, above and below Earth (2 words)

9   A storm with winds of 75 miles per hour or greater and lots of rain

13  Water changing from liquid to gas

14  Summer, autumn, winter, and spring

16  Degree of hotness or coldness

21  We get our heat and light from it

22  An electric discharge in the atmosphere

24  Moving air

25  Rain, snow, sleet, and hail are forms

**DOWN**

1   Average weather over a period of time

2   State of the atmosphere at a given time and place

4   A severe winter storm

5   The amount of water vapor in the air

7   Water changing from gas to liquid

8   A boundary between two air masses

9   Precipitation made up of layers of ice and snow

10  Precipitation in the form of drops of water

11  A cloud that is very close to the ground

12  A region that gets very little rain

15  A light wind

17  Loud sound that we hear before we see lightning

18  Imaginary line around the center of the Earth

19  Condensed vapor in the atmosphere; cumulus is one type

20  Precipitation in the form of ice crystals

23  Dangerous storm with a funnel-shaped cloud

# Weather and Climate

# Science Terms

## ACROSS

2    Place in an ecosystem where an organism usually lives

3    To move back and forth quickly

4    A series of organisms that depend on one another for food (2 words)

6    Community of the living and nonliving things that share an environment

8    A tool used to measure the amount of heat energy in something

9    One of the four natural divisions of the year

10  Stages in an organism's life (2 words)

13  A measure of the amount of energy in something

17  Vertebrates have one; invertebrates do not

19  Visible electromagnetic radiation

20  A balance ___ is used to measure mass

21  A very large group of stars; Milky Way is one

22  A tasteless, transparent, odorless liquid

## DOWN

1    To use our senses to identify or learn about something

2    Passage of traits from parents to offspring

5    The sun and all the objects that orbit it

7    What like poles of a magnet do

9    Person who studies or investigates a field of science

11  To place things that have similar properties into groups

12  A planet's satellite

14  A combination of 2 or more substances

15  Data collected as a result of observations and experiments

16  Everything that exists

18  What unlike poles of a magnet do

# Science Terms

# Animal Babies Word Search

```
F  P  U  P  P  Y  Q  T  K  V  I
W  L  K  I  T  T  E  N  G  P  O
K  A  A  Q  Y  O  E  N  C  P  F
G  I  U  C  Y  W  I  L  W  X  C
N  S  D  E  H  L  L  E  W  A  V
I  D  O  E  K  B  P  A  I  O  F
L  J  L  C  C  J  I  G  O  S  L
S  P  U  H  C  C  G  L  S  F  M
O  D  I  D  V  P  L  E  A  C  H
G  C  T  L  T  F  E  T  U  M  M
K  D  F  U  Z  K  T  B  T  C  B
```

| | |
|---|---|
| CALF | JOEY |
| CHICK | KID |
| CUB | KITTEN |
| DUCKLING | LAMB |
| EAGLET | OWLET |
| FAWN | PIGLET |
| FOAL | PUPPY |
| GOSLING | WHELP |

# Body Parts Word Search

```
M L I M S U S R A E M D
R X P V H A N D S R S A
O T K O J O I N T S M E
B R A I N O S E B R R H
O A V S R X Z G O M A F
O E G S E T E A N K L E
E H H N D V S K E L S S
V Y R T L F R C S E G K
S J E X U C K E G G N U
K Z N S O O Y N N S U L
I Q E P H A M K S T L L
N Y M U S C L E S G L P
```

| | |
|---|---|
| ANKLES | LEGS |
| ARMS | LUNGS |
| BONES | MOUTH |
| BRAIN | MUSCLES |
| EARS | NECK |
| EYES | NERVES |
| HANDS | NOSE |
| HEAD | SHOULDER |
| HEART | SKIN |
| JOINTS | SKULL |

# Hidden Words: Our Five Senses

There are five main senses: sight, hearing, touch, taste, and smell.
A word relating to one of these senses is hidden in each sentence.
Use the clues to find the hidden words. Underline those words.

1.   Mom adds parsley every time she makes soup.
CLUE: One of the organs responsible for sight

2.   Sari made a beautiful heart pendant for her mother.
CLUE: One of the organs responsible for hearing

3.   Dad said to Eton, "Guests are coming, so pick up your things."
CLUE: Organ needed for taste

4.   Jessie wore a mask into the party.
CLUE: Organ needed for touch

5.   Emma ordered a cappuccino, Seth ordered a soda, and Dad ordered milk.
CLUE: Organ needed for sense of smell

6.   When Kelly let out a sigh, Tom turned around to look at her.
CLUE: Sense involving seeing

7.   Staci was working on her science project and said, "Pass me llama pictures, please."
CLUE: Sense involving perceiving odors

8.   In each ear Ingrid wore a large, dangling earring.
CLUE: Sense that involves our ears

9.   For dinner I ate pasta, steak, and carrots.
CLUE: Sense that makes me love chocolate ice cream

10.  Mr. Intou chatted with his son's teacher.
CLUE: Sense that allows us to feel if something is smooth or rough

# Solutions*

**\*Optional Lists of Answers**

Alphabetical lists of the answers are provided. These may be used to help solve the puzzle from the beginning, to assist those having difficulty, or not at all.

# Animal Characteristics

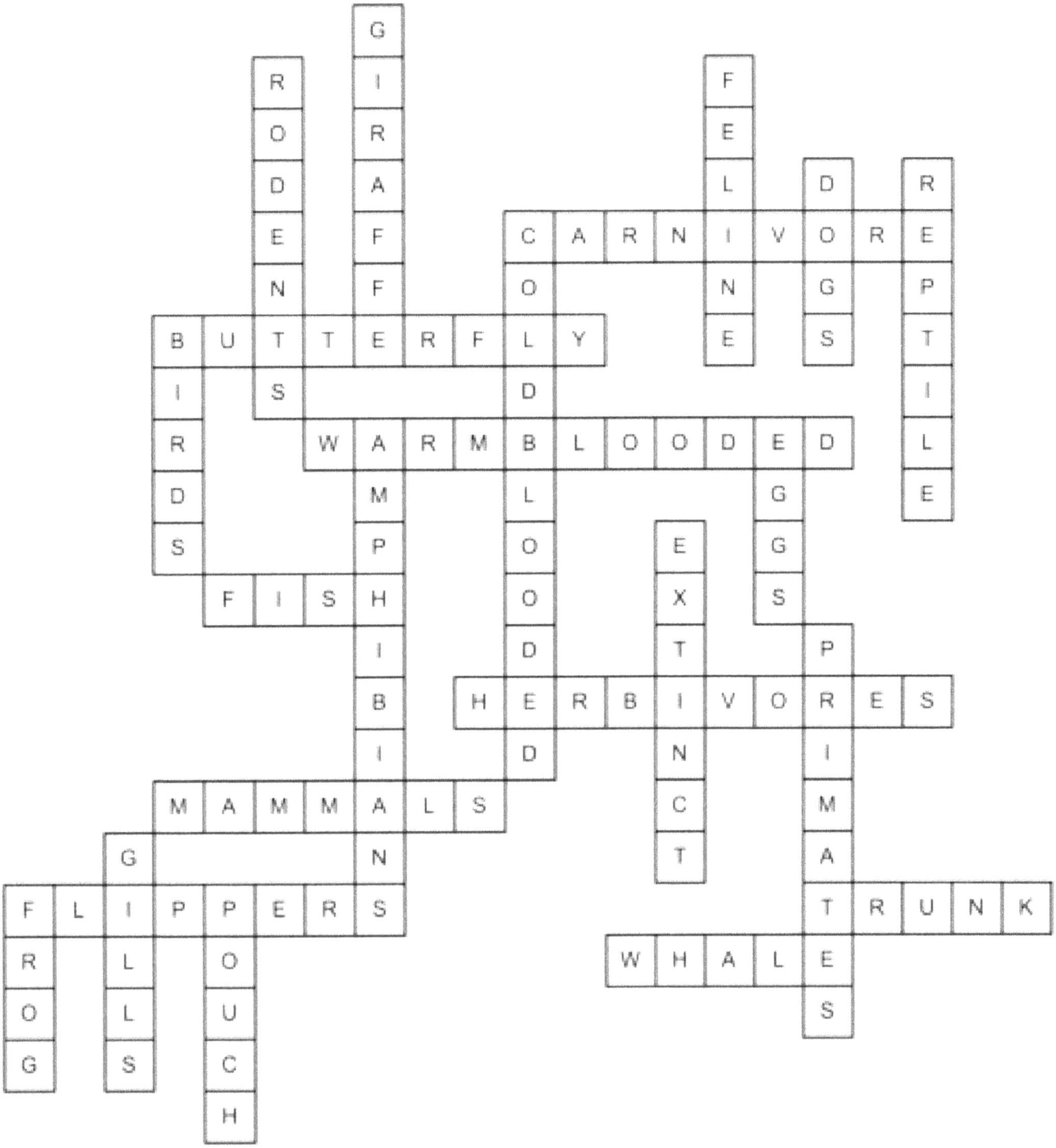
GIRAFFE
RODENTS
FELINE
DOGS
REPTILE
CARNIVORE
COLD
BUTTERFLY
BIRDS
WARMBLOODED
EGGS
AMPHIBIANS
BLOOD
EXTINCT
FISH
PRIMATES
HERBIVORES
MAMMALS
GILLS
FLIPPERS
FROG
POUCH
TRUNK
WHALE

# Animal Families

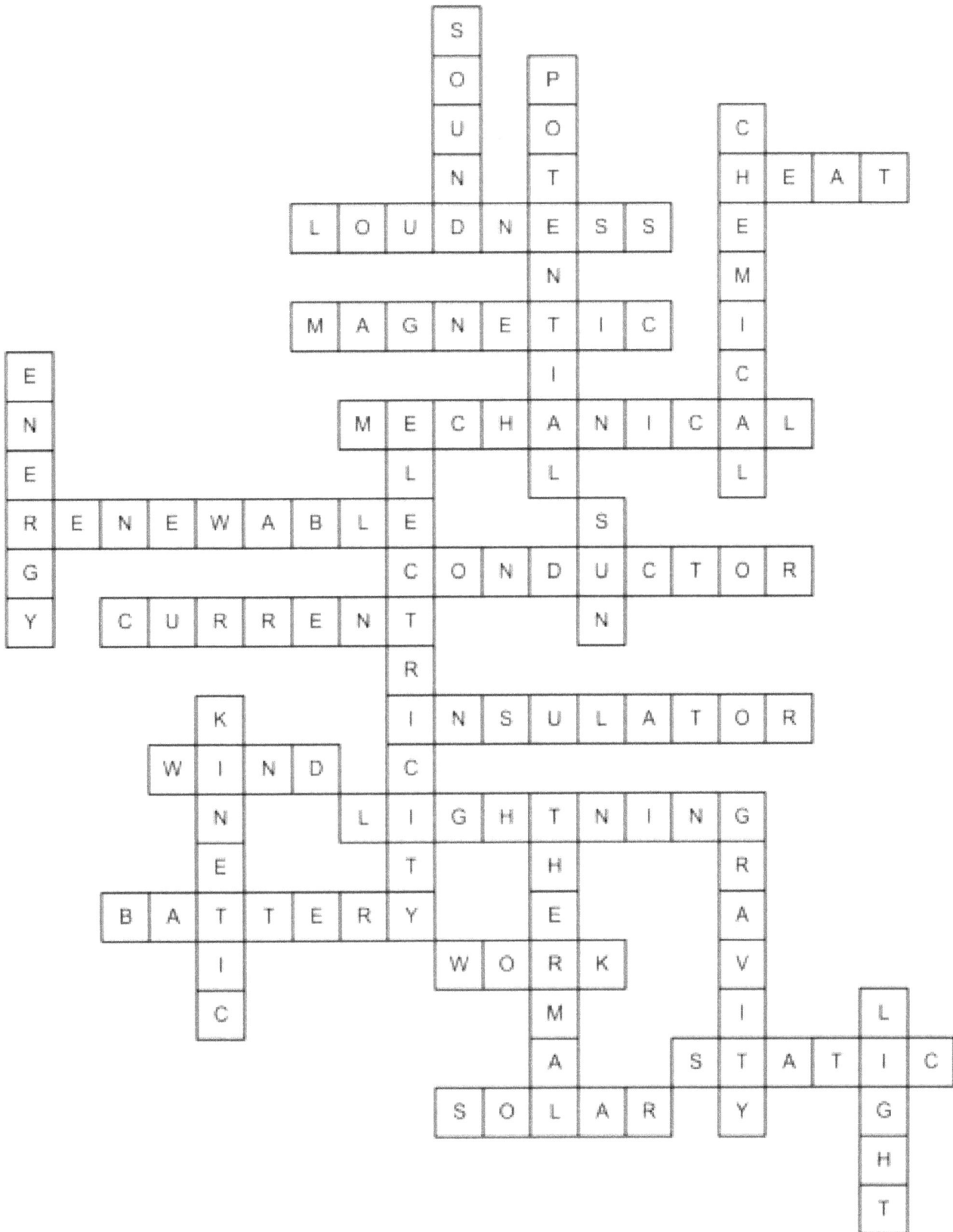

# Energy

```
            S
            O            P
            U            O                        C
            N            T            H  E  A  T
L  O  U  D  N  E  S  S   E                        E
                         N                        M
M  A  G  N  E  T  I  C   T                        I
                         I                        C
E           M  E  C  H  A  N  I  C  A  L
N           L            L                     L
E           L
R  E  N  E  W  A  B  L  E            S
G           C  O  N  D  U  C  T  O  R
Y     C  U  R  R  E  N  T            N
            R
      K     I  N  S  U  L  A  T  O  R
W  I  N  D  C
      N     L  I  G  H  T  N  I  N  G
      E     T            H            R
B  A  T  T  E  R  Y      E            A
      I     W  O  R  K   R            V
      C                  M            I     L
                         A   S  T  A  T  I  C
            S  O  L  A  R    Y            G
                                         H
                                         T
```

# Force and Motion

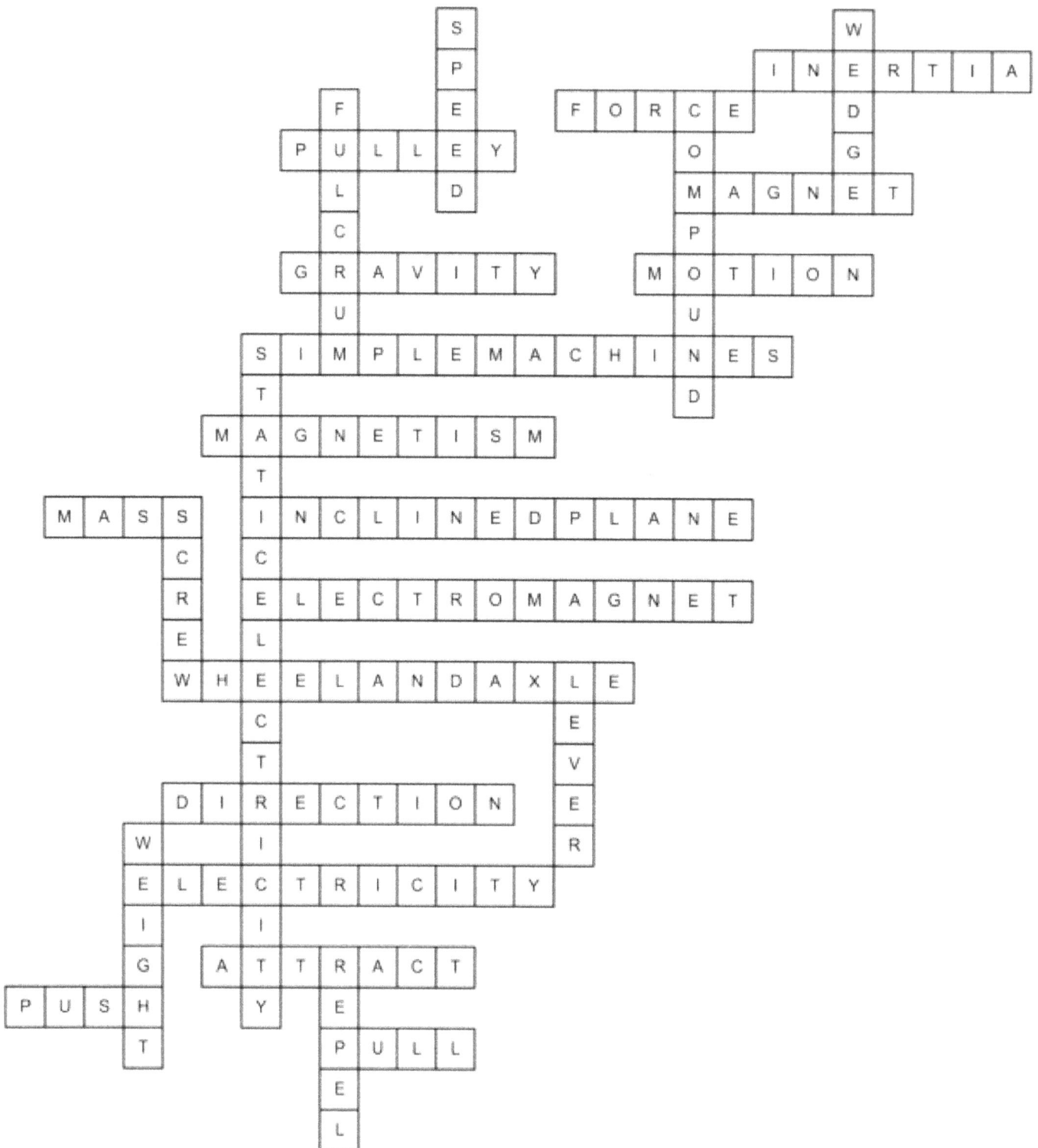

A completed crossword puzzle with the following words:

- SPEED (vertical)
- FULCRUM (vertical, spelled F-U-L-C-R-U-M)
- PULLEY
- GRAVITY
- SIMPLE MACHINES
- MAGNETISM
- MASS
- INCLINED PLANE
- ELECTROMAGNET
- WHEEL AND AXLE
- DIRECTION
- ELECTRICITY
- ATTRACT
- PUSH
- PULL
- FORCE
- INERTIA
- WEDGE
- MAGNET
- MOTION
- COMPOUND
- STATIC
- SCREW
- FRICTION
- WEIGHT
- LEVER
- REPEL

# The Human Body

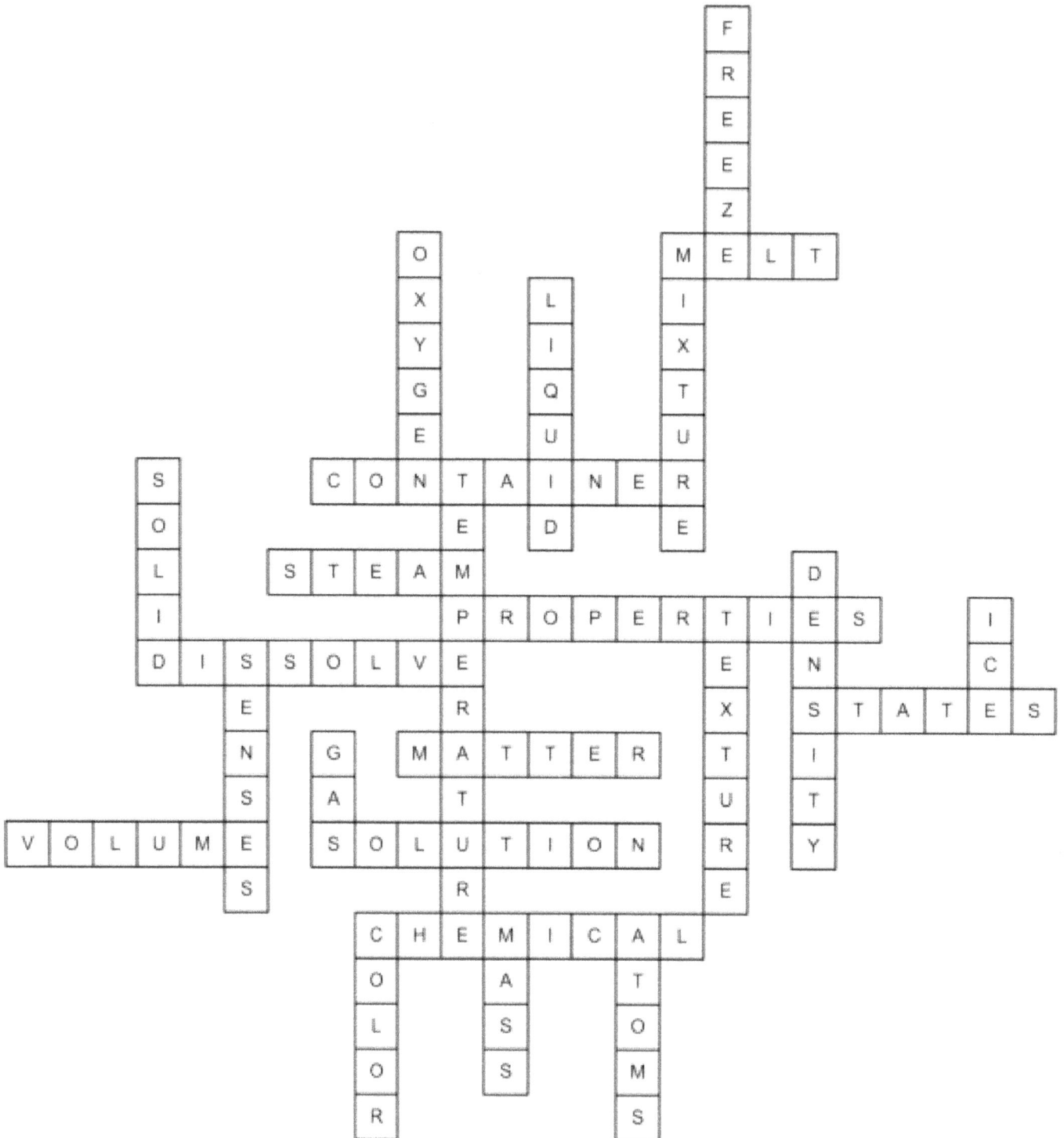

# Matter

# Planet Earth

OCEAN
CORE
GLACIER
MAGMA
FAULT
PLANET
MOUNTAIN
RIVER
MINERAL
VOLCANO
EARTHQUAKE
WATERCYCLE
STAR
SOIL
CONTINENT
MANTLE
SUN
CRUST
TIDE
LANDSLIDE
WEATHERING
EROSION
ICEBERG
ROCKS
MOON

# Plants

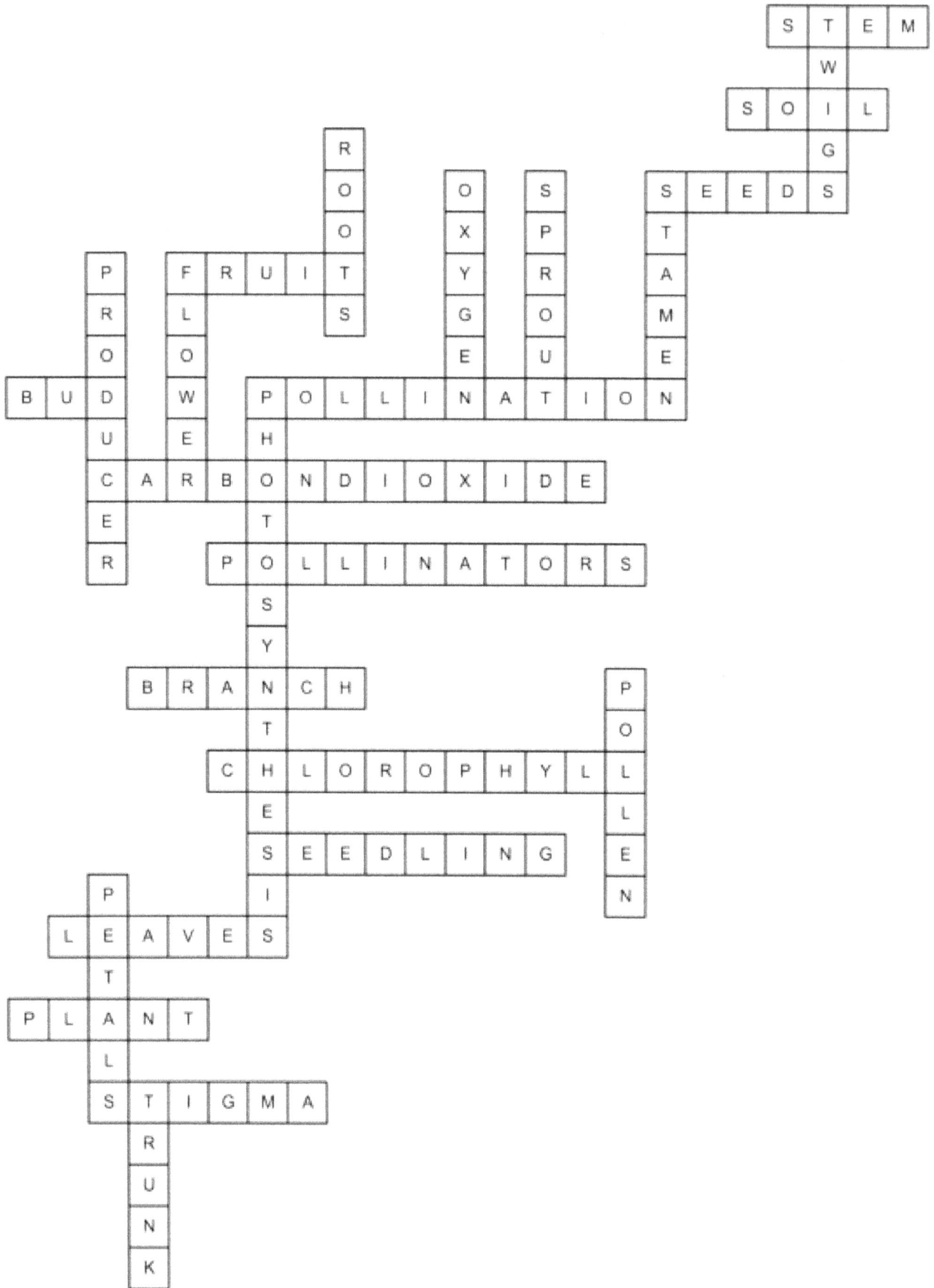

# Weather and Climate

# Science Terms

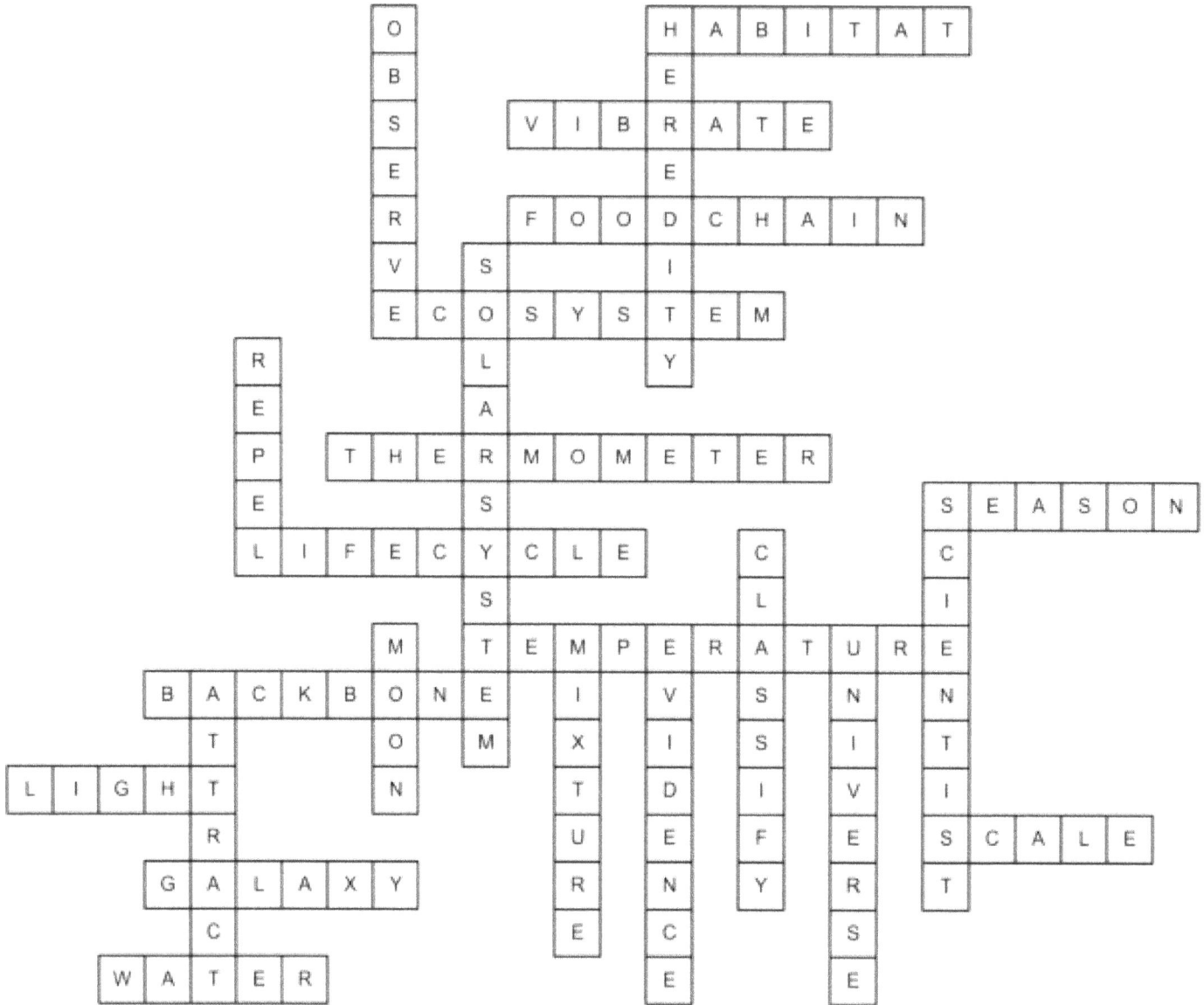

A crossword puzzle grid containing the following words:

OBSERVE, HABITAT, VIBRATE, FOODCHAIN, ECOSYSTEM, THERMOMETER, SEASON, LIFECYCLE, REPTILE, TEMPERATURE, BACKBONE, LIGHT, GALAXY, WATER, CLASSIFY, MIXTURE, EVIDENCE, UNIVERSE, CLIMATE, SCALE, ATTRACT, ELEMENT, CLIMATE

Down words include: OBSERVE, HEREE, SOLARSYSTEM, DIVIDE, CLASSIFY, UNIVERSE, CLIMATE, MIXTURE, EVIDENCE, ATTRACT

# Animal Babies Word Search

|    | 1 | 2 | 3 | 4 | 5 | 6 | 7 | 8 | 9 | 10 | 11 |
|----|---|---|---|---|---|---|---|---|---|----|----|
| 1  | F | P | U | P | P | Y | Q | T | K | V  | I  |
| 2  | W | L | K | I | T | T | E | N | G | P  | O  |
| 3  | K | A | A | Q | Y | O | E | N | C | P  | F  |
| 4  | G | I | U | C | Y | W | I | L | W | X  | C  |
| 5  | N | S | D | E | H | L | L | E | W | A  | V  |
| 6  | I | D | O | E | K | B | P | A | I | O  | F  |
| 7  | L | J | L | C | C | J | I | G | O | S  | L  |
| 8  | S | P | U | H | C | C | G | L | S | F  | M  |
| 9  | O | D | I | D | V | P | L | E | A | C  | H  |
| 10 | G | C | T | L | T | F | E | T | U | M  | M  |
| 11 | K | D | F | U | Z | K | T | B | T | C  | B  |

## Word Search Answer/Hints
### The words below are listed with their starting row and column.

CALF  4:4          JOEY  7:2

CHICK  7:5         KID  3:1

CUB  9:10          KITTEN  2:3

DUCKLING  9:2      LAMB  8:8

EAGLET  5:8        OWLET  6:10

FAWN  6:11         PIGLET  6:7

FOAL  8:10         PUPPY  1:2

GOSLING  10:1      WHELP  4:6

# Body Parts Word Search

|    | 1 | 2 | 3 | 4 | 5 | 6 | 7 | 8 | 9 | 10 | 11 | 12 |
|----|---|---|---|---|---|---|---|---|---|----|----|----|
| 1  | M | L | I | M | S | U | S | R | A | E  | M  | D  |
| 2  | R | X | P | V | H | A | N | D | S | R  | S  | A  |
| 3  | O | T | K | O | J | O | I | N | T | S  | M  | E  |
| 4  | B | R | A | I | N | O | S | E | B | R  | R  | H  |
| 5  | O | A | V | S | R | X | Z | G | O | M  | A  | F  |
| 6  | O | E | G | S | E | T | E | A | N | K  | L  | E  |
| 7  | E | H | H | N | D | V | S | K | E | L  | S  | S  |
| 8  | V | Y | R | T | L | F | R | C | S | E  | G  | K  |
| 9  | S | J | E | X | U | C | K | E | G | G  | N  | U  |
| 10 | K | Z | N | S | O | O | Y | N | N | S  | U  | L  |
| 11 | I | Q | E | P | H | A | M | K | S | T  | L  | L  |
| 12 | N | Y | M | U | S | C | L | E | S | G  | L  | P  |

## Word Search Answer/Hints
The words below are listed with their starting row and column.

ANKLES  6:8

ARMS  5:11

BONES  4:9

BRAIN  4:1

EARS  1:10

EYES  7:1

HANDS  2:5

HEAD  4:12

HEART  7:2

JOINTS  3:5

LEGS  7:10

LUNGS  11:11

MOUTH  11:7

MUSCLES  12:3

NECK  10:8

NERVES  10:9

NOSE  4:5

SHOULDER  12:5

SKIN  9:1

SKULL:  7:12

# Hidden Words: Our Five Senses

There are five main senses: sight, hearing, touch, taste, and smell.
A word relating to one of these senses is hidden in each sentence.
Use the clues to find the hidden words. Underline those words.

1.   Mom adds pars<u>ley e</u>very time she makes soup.
CLUE: One of the organs responsible for sight

2.   Sari made a beautiful h<u>ear</u>t pendant for her mother.
CLUE: One of the organs responsible for hearing

3.   Dad said to E<u>ton, "Gue</u>sts are coming, so pick up your things."
CLUE: Organ needed for taste

4.   Jessie wore a ma<u>sk in</u>to the party.
CLUE: Organ needed for touch

5.   Emma ordered a cappucci<u>no, Se</u>th ordered a soda, and Dad ordered milk.
CLUE: Organ needed for sense of smell

6.   When Kelly let out a <u>sigh, T</u>om turned around to look at her.
CLUE: Sense involving seeing

7.   Staci was working on her science project and said, "Pas<u>s me l</u>lama pictures, please."
CLUE: Sense involving perceiving odors

8.    In eac<u>h ear In</u>grid wore a large, dangling earring.
CLUE: Sense that involves our ears

9.   For dinner I ate pa<u>sta, ste</u>ak, and carrots.
CLUE: Sense that makes me love chocolate ice cream

10.  Mr. In<u>tou ch</u>atted with his son's teacher.
CLUE: Sense that allows us to feel if something is smooth or rough

# Optional Lists of Words and Terms

These lists are provided for your convenience should you choose to use them.

## Animal Characteristics

amphibians    birds    butterfly    carnivore    cold-blooded

dogs    eggs    extinct    feline    fish    flippers    frog

gills    giraffe    herbivores    mammals    pouch    primates

reptile    rodents    trunk    warm-blooded    whale

## Animal Families

calf    colt    cub    doe    drake    ewe

filly    flock    foal    gosling    herd    kangaroo

kid    kit    kitten    lamb    mare    mob

pride    puppy    school    stallion    troop

## Energy

battery    chemical    conductor    current    electricity    energy

gravity    heat    insulator    kinetic    light    lightning

loudness    magnetic    mechanical    potential    renewable

solar    sound    static    sun    thermal    wind    work

## Force and Motion

attract    compound    direction    electricity    electromagnet    force    fulcrum

gravity    inclined plane    inertia    lever    simple machines    magnet

magnetism    mass    motion    pull    pulley    push    repel    screw

speed    static electricity    wedge    weight    wheel and axle

## The Human Body

ankle    arm    blood    bones    brain    cell    chest

elbow    face    head    heart    joint    knee    lungs    neck

nerves    senses    shin    shoulder    skeletal    skin    skull

spinal cord    system    thigh    tissue    waist    wrist

## Matter

atoms      chemical      color      container      density      dissolve

freeze      gas      ice      liquid      mass      matter      melt

mixture      oxygen      properties      senses      solid      solution

states      steam      temperature      texture      volume

## Planet Earth

continent      core      crust      earthquake      erosion      fault

glacier      iceberg      landslide      magma      mantle      minerals

moon      mountain      ocean      planet      river      rocks

soil      star      sun      tide      volcano      water cycle      weathering

## Plants

branch      bud      carbon dioxide      chlorophyll      flower      fruit

leaves      oxygen      petals      photosynthesis      plant      pollen

pollination      pollinators      roots      producer      seedling      seeds

soil      sprout      stamen      stem      stigma      trunk      twigs

## Weather and Climate

blizzard      breeze      climate      cloud      condensation      desert

equator      evaporation      fog      front      hail      humidity      hurricane

lightning      precipitation      rain      seasons      snowstorm      sun

temperature      thunder      tornado      water cycle      weather      wind

## Science Terms

attract      backbone      classify      ecosystem      evidence

food chain      galaxy      habitat      heredity      life cycle      light      mixture

moon      observe      repel      scale      scientist      solar system

temperature      thermometer      universe      vibrate      water

www.ingramcontent.com/pod-product-compliance
Lightning Source LLC
Chambersburg PA
CBHW051428200326
41520CB00023B/7397